Let's learn about military work today!

Service members are brave, dedicated, and help keep us safe.

Military work starts with training. They practice skills again and again.

Training helps them learn how to do their jobs well.

Military work is about teamwork.

No one works alone.

They help each other
and work side by side.

Military workers stay ready.
They gear up and make plans.

Being ready lets them support others when it matters most.

One important job is protection.

Service members watch over places and people.

Their work helps keep communities safe.

Helping others is
an important part of the job.

They lend a hand when someone needs it.

Sometimes military work
means traveling to new places
to help people.

Service members are brave and flexible wherever they go.

Military work happens
in many places.

On land. On water. In the air.

The job may change, but the purpose stays the same.

Military workers trust each other.
They depend on their teammates.

Trust helps them work safely and do their jobs well.

Service members take pride in their work.

They practice carefully every day.

Their work matters to many people.

Military work can be tiring.
Some days are long.

Even then, they keep going
and do their best.

After working hard,
it's time to head home.

Rest and family are important too.

Military work matters.

And the families who support them matter too.

Made in United States
Troutdale, OR
10/30/2026

56609885R00017